DISCARD

# SPACE AGE FIRE FIGHTERS
## New Weapons in the Fireman's Arsenal

## by C. B. COLBY

Coward, McCann & Geoghegan, Inc.     New York

# Contents

| | |
|---|---|
| Space-Age Fire Fighting | 3 |
| 1000 GPM Pumper | 4 |
| Cab-Forward Mack | 6 |
| Specially Equipped Pumper | 8 |
| New York's Superpumper System | 10 |
| Superpumper in Action | 12 |
| Foam, the Fire Killer | 14 |
| Meet the Tele-Squrt | 16 |
| The Snorkel | 18 |
| Versatile Tower Ladders | 20 |
| Old Reliable Aerial Ladders | 22 |
| Fireboats in Action | 24 |
| New York Fire Department Scuba Firemen | 26 |
| "Space-Powered" Jet-Axe | 28 |
| Rescue Rocket Torch | 30 |
| Multipurpose Saw | 31 |
| Labor-Saving Water-Vak | 32 |
| Lifesaving "Teepee" | 33 |
| Space-Age Fire-Safe Fabrics | 34 |
| Space-Program Fire Fighters | 36 |
| Fireproof and Germproof | 37 |
| Machines Against Forest Fires | 38 |
| Crash Pads for Firemen's Tools | 40 |
| Science vs. Flames | 41 |
| Flying "Fire Trucks" | 42 |
| Firemen in Jumpsuits | 44 |
| Down, and Up a Tree | 46 |
| What It's All About? | 48 |

Photo Credits: Fire Department, New York City, 1, 2, 10, 12, 13, 20, 21, 22, 23, 24, 25, 26, 27, 48. United States Forest Service, 33, 38, 39, 40, 41, 42, 43, 44, 45, 46, 47. National Aeronautics and Space Administration, full-color cover and 34, 35, 36, 37. United States Coast Guard, 15, lower. Snorkel Fire Equipment Co., 16, 17, 18, 19. The Port of New York Authority, 14. Explosive Technology, a subsidiary of OEA, Inc., 28, 29. Mack Trucks, Inc., 4, 5, 6, 7, 8, 9. Mine Safety Appliances Co., 15, upper. United Technology Center, 30. All others by the author.

Second Impression

© 1973 by C. B. Colby     All rights reserved     Printed in the United States of America
SBN: GB-698-30531-0
SBN: TR-698-20279-1

Library of Congress Catalog Card Number: 73-82034

# Space Age Fire Fighters

Since my first book on fire fighting was published in 1954, not only has fire-fighting equipment changed but fires themselves. The use of new materials, chemicals, and architectural design has brought about changes in the way fires ignite, burn, travel, and have to be extinguished.

Many of the new plastics emit dangerous fumes when burned, some new fabrics are potentially dangerous when exposed to high temperatures, and exotic new chemical combinations become hazardous when stored under certain conditions. New protective clothing, fire-fighting chemicals, equipment, and techniques have had to be developed to meet the new challenge.

New aerial fire-fighting platforms bring men and water into a more effective position for fighting flames aloft. Tremendously powerful pumping engines have been developed to pour more water onto a fire than ever before. Even space-age rockets have been miniaturized and refined into powerful cutting torches for rescue work.

High-speed gasoline saws equipped with new blades can slice through concrete and metal walls, floors and roofs to ventilate a fire or rescue a victim, while a space-age product, the Jet-Axe, can blast a hole of predetermined size and shape through a metal door, concrete wall or other barrier.

Firemen wearing clothing of new fireproof and heat-reflecting material can literally walk through flames. And new fire-retardant paints and fabrics will someday make furnishings and interiors as safe from burning as a pail of water. New fire-fighting foams and vapors can smother a gasoline or oil fire in seconds. A radical new fire-extinguishing vapor by DuPont, called Halon 1301, can snuff out a fire with an invisible blanket, while furnishing enough oxygen to permit trapped persons to breath and escape even while the fire around them is being extinguished!

New and stronger hoses have been engineered; and new, more effective nozzles to direct water onto the flames have been designed for them. Labor-saving devices of all kinds help the cleanup squads make dwellings livable after a fire; window and door openings can be sealed with tough plastic sheets and roof openings can be closed against the weather, while powerful water-inhaling vacuum pumps suck water from floors and basements.

Powerful multistream fireboats, snorkels, Squrts, and ladder towers, as well as such fantastic weapons as the New York City Fire Department's superpumper, supertender and its satellites make today's fire fighters a match for any blaze.

With today's emphasis on environmental protection, the forest-fire fighter takes on an even greater importance. New weapons give him potent means to battle wilderness fires more effectively. Helicopters can carry kingsized buckets of water to dump on hot spots. And tanker planes can make sensational drops of fire-retardant chemicals from treetop level onto the flaming forest.

Infrared instruments carried by aircraft overhead can peer through dense clouds of thick smoke to pinpoint the fire lines and hot spots, although the ground and flames below are invisible to the men aboard the aircraft. New packaging can safely and accurately deliver tools to the fire fighters.

On the following pages you will see many of these new weapons used by firemen of the space age, whether fighting a fire in a private dwelling, a city skyscraper or a timbered wilderness. Each type of fire requires a special set of weapons, the know-how to use them efficiently and well-trained firemen to make an attack effective. Perhaps this book will encourage you to visit your own fire department and get to know the men and their equipment, a worthwhile and exciting experience for the whole family.

I would like to thank many people for their kindness and cooperation in helping me obtain these exceptional photos. A few of them were also used in my recent book, *Space Age Spinoffs*; they are the only available photographs of these brand-new space-age fire-fighting weapons.

I would like to express particular thanks to the following fine folks of the New York City Fire Department; Commissioner Robert O. Lowery; Deputy Assistant Chief, Herbert F. Whyte; Captain John F. Dahl, Jr., of the Community Relations Bureau; and Lieutenant Edward Regulinsky and Fireman Cecil Maloney of Satellite 1.

My thanks also to the many manufacturers for their help and to John E. McLeaish, chief of the Public Information Office, Public Affairs Office of the Manned Spacecraft Center, NASA, Houston, Texas, and to J. L. Hickman, Forest Service, U.S. Department of Agriculture, Washington, D.C. To these, my appreciation for helping make this book a pleasant project.

C. B. Colby

# 1000 GPM Pumper

This is a fine example of a tilt-cab pumper. It carries a crew of five men, two in the front seat and three behind, facing to the rear. It can pump at the rate of 1,000 gallons of water a minute; and it carries a 500-gallon booster tank with preconnected hoses on reels. This means that as soon as it reaches the fire scene, it can start pumping water onto the blaze without waiting to hitch to a hydrant, while its heavier lines (hoses) are being connected. Often the 500-gallon tank will hold a fire in check; it may even extinguish the fire without using the larger lines. Such trucks are excellent for grass and small brush fires, for washing oil and gasoline from a highway after an accident, and for fighting other fires that occur far away from hydrants. The sections of big stiff hose on the sides are called "hard-suction pipes." They are used to draft (suck) water from ponds, swimming pools or streams when there are no hydrants nearby. A screen over the end in the water keeps out debris, pebbles, fish, frogs and other pump-clogging material. On the opposite page is a closeup of a pump control panel and rear-facing seats. Extra men can ride on the rear of the truck if necessary. The entire cab tilts forward to permit work on the big diesel engine under the front seats.

## Cab-Forward Mack

In this model the engine is behind the cab where there are also seats for extra men. Note the hard-suction pipes on the left side and the small roof ladder on the right side. The circular object in the center front is the siren. It also has air horns on the cab roof. This fine pumper also has a 500-gallon booster tank and two preconnected lines for a fast attack at the fire scene. The two spools on each side of the framework in front of the two reels of preconnected hose are rollers. Rollers make it easier to pull these two booster lines when they must be pulled to the side instead of straight forward off the reels. The two hoses can be rewound either manually or by using power. The photo opposite is a closeup of the pump control panel, with water-supply gauges and other instruments that inform the driver and operator of the condition of the equipment and the supply of water in the booster tank. Several hundred feet of hose can be carried in the center of these pumpers. The hoses are usually two and one half inches in diameter and are made of canvas or nylon fabric. The truck engine also drives the pumps and can be controlled from this panel.

## Specially Equipped Pumper

Every department engaged in fire fighting has certain requirements for the equipment they order. Some of the requirements are unusual. Some have a plow on the front to enable their "brush busters" to get through light saplings to reach grass fires in rural areas. Others, such as this pumper designed by Mack, have protective roofs over all the crew areas as shields against falling embers, debris, and missiles hurled from rooftops. This pumper has a completely enclosed crew cab and a deck-mounted deluge gun (high-pressure nozzle) over the control panel. Thanks to its uniquely curved pipe mount, this nozzle can be turned in any direction by its control wheel or locked in position to send a steady stream required at one spot. It also has a booster tank and a power-driven preconnected hose reel on the back of the truck. Three sections of hard-suction hose are carried as well as an extension ladder. On the opposite page is a closeup of the control panel with its many dials, valves and hose connections. The big connection with two stub handles is where the hard-suction pipe is connected.

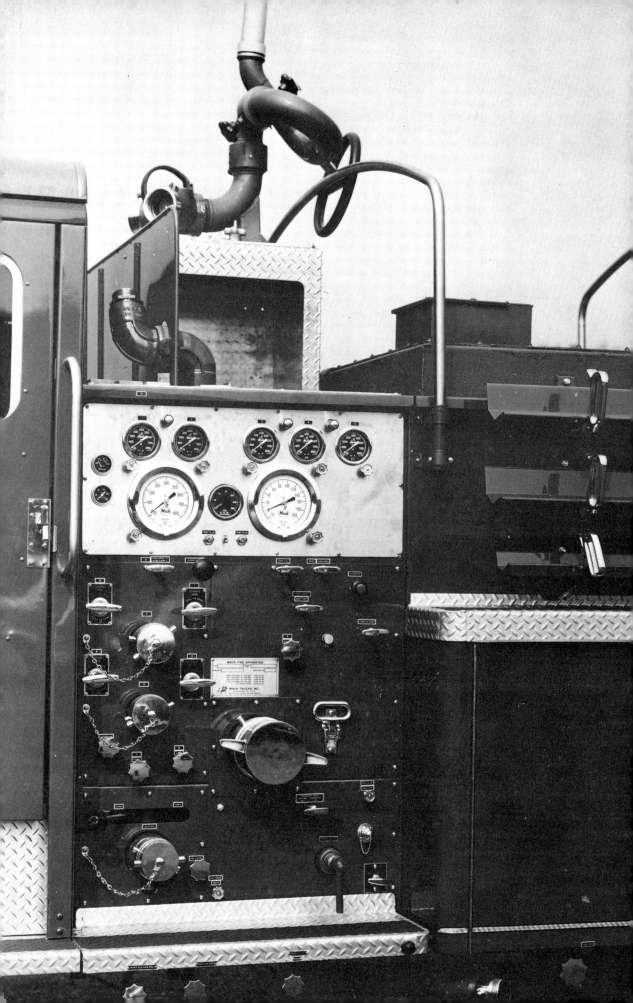

# New York's Superpumper System

In a city as large as New York, fighting a huge fire can reduce the city's water pressure to a dangerously low level. To counteract this, the city had this superpumper designed and built. It is actually a mobile pumping station with fantastic power. The main pumping unit is shown above. The superpumper system consists of the main unit plus four other units: a supertender, which carries approximately 2,000 feet of hose, nozzle fittings and other equipment used by the superpumper, and three satellite units, each with another 2,000 feet of hose, various types of nozzle tips but no pumps. The superpumper can pump water over a mile and has enough power to smash doors, barricades and even walls with its high-pressure streams. The superpumper is mainly used to provide large volumes of water, and streams of water up to 600 feet long where a fire is too hot to approach close enough to use conventional streams. It can also be used to pump river water to a fire away from the dock area, if other water is short or pressure is too low. For example, at one fire alone, 5,000,000 gallons of river water were supplied by this pumper unit. It takes four hydrants to supply this monster, while usually two or more normal-sized pumpers can operate from a single hydrant. The superpumper can pump nearly 10,000 gallons per minute; it could fill a twenty foot by forty foot swimming pool in five minutes or pump more than its own weight of 68,500 pounds in water in one minute. The opposite page shows (above) the supertender and (below) one of the satellite units of this superpumper system.

## Superpumper in Action

Above is a dramatic photo of the superpumper in action during a winter storm. This is the worst possible time for a fire because ice soon coats the streets, ladders, tools, personnel gear and even the fire fighter's mustache. This is the pumping end of the unit where water is received from hydrants or the river and pumped out to the supertender or satellite units. The gigantic DeLaval pump has eight outlets and can pump at the rate of 4,400 gallons per minute at 700 psi (pounds per square inch) or 8,800 gallons per minute at 350 psi. At a lower psi it can pump as much as 10,000 gallons a minute. The superpumper is over forty-three feet long and eight feet wide. The engine is a Napier Delta with eighteen cylinders and is started by air pressure at 450 psi. It is rated at 2,400 horsepower and carries 400 gallons of fuel. It uses 140 gallons of fuel an hour at maximum capacity. On the opposite page is a closeup of the high-pressure nozzle of one of the satellite units in action. It seems to be dangerously close to the men in the tower ladder, but they are safely out of the way of the stream.

# Foam, the Fire Killer

A fire needs three things to keep going: fuel, heat and oxygen. If you take away any one of the three, the fire goes out. If you remove the fuel—for example, if you shut off the gas in a stove—the fire goes out. If you cool it down with enough water, the combustion stops and the fire goes out. If you remove the oxygen—for example, if you put a cover over a can of burning fuel or a blanket over a person with clothing on fire—you cut off the oxygen and the fire goes out. The best way to extinguish many types of fires, such as gasoline, electrical or oil fires, is by smothering them. Foam is the best way to do this, and here are some examples of how foam is carried to the fire scene. Above is a special foam truck called the Jumbo E-77, used at Kennedy International Airport in New York and Newark International Airport in New Jersey. This truck can spread 2,000 gallons of foam a minute from its two remote-controlled turrets mounted on the top deck. The turrets can also be operated manually, in case of automatic-control failure. A crew of two can operate this emergency vehicle; one to drive and operate the front turret and a second to operate the rear turret. It can carry 7,000 gallons of water and 700 gallons of foam-making chemical and can spread the 7,700 gallons of fire-smothering foam over a burning aircraft in less than four minutes. On the opposite page (top) is a portable foam nozzle called the Mini-X Foamaker made by Mine Safety Appliances Company. It can be attached to a regular fire hose and used by one fireman. It uses a one and a half inch line and foam concentrate and is ideal for fighting gasoline and oil fires from a distance of ten or twelve feet. The lower photo shows a U.S. Coast Guard fire fighter in a protective outfit using foam to smother a shipboard fire during a training exercise.

# Meet the "Tele-Squrt"

This ingenious device combines an aerial ladder with a water tower so that water can be directed into burning buildings from some distance away. This combination ladder and tower can be extended vertically or horizontally for fifty feet. It is made by the Snorkel Fire Equipment Company and can be designed into new trucks or added to those already in use. The stream can be directed from the truck itself or from the top of the unit by a man at either station. The low ladder rails give protection during a rescue. It can be raised or lowered in less than a minute, and with the boom in any normal position it can easily and safely handle streams of 1,000 gallons per minute. The top photo shows the truck with the Tele-Squrt retracted and lowered for travel. The lower photo shows the unit with the boom extended horizontally and the stream in action. On the opposite page the unit is being tested as a water tower. Note the outriggers bracing the truck against tipping in both action photos. The Tele-Squrt is a versatile fire-fighting tool.

# The "Snorkel"

The familiar "cherry picker" device used by power and phone companies to reach the tops of utility poles is also used by fire departments and is generally called a snorkel. It is made by the same company that produced the Tele-Squrt on the preceding page. It is designed with a folding bend-in-the-middle boom—like someone's arm, with a platform where the hand would be and a joint where the elbow is. It is shown (above) folded in the middle and lying flat across the top of the truck on which it is mounted. The low silhouette is excellent, and the clean lines of the folded boom show fine engineering. The photo on the opposite page shows the snorkel in action in an elevated position. The platform can carry two or more persons, but a crew of two is usual. Full controls are installed in the platform, but the boom can be operated from the truck itself in case the men on the platform are unable to function. The deck gun (nozzle) can be directed in any direction; and various types of nozzles, including a foam nozzle, can be used. The platform is equipped with gates in the sides so that rescues can be made through the sides rather than over the railings. The snorkel design permits fire fighters to reach roof fires that are blocked to other equipment by high-tension wires or building setbacks—a highly maneuverable innovation in fire-fighting equipment.

## Versatile Tower Ladders

One of the most valuable types of fire-fighting equipment is the telescoping tower ladder shown above and on the opposite page. These tower ladders can reach up or out to seventy-five feet to direct streams of water or foam into the heart of the fire, over utility wires, under piers or even into structures themselves if necessary. As many as three men can mount the attack on the fire from the platform, and the ladder sections that extend with the boom sections permit personnel changes or trapped citizens to escape in an emergency, without retracting the boom itself. The dramatic photo above shows a tower ladder being used at low level; and the photo on the opposite page shows two of these ladders extended to reach higher floors. One of the advantages of the tower ladders is their great strength, which enables them to carry as many fire victims as can fit onto the platform. Outriggers are used to brace the truck against tipping. These tower ladders are made by Mack and are called aerialscopes or "the big reach." They can be used as water towers, using a straight stream or fog nozzle; as lifts for personnel, hose or equipment; as exterior elevators for window or roof rescues; or as elevated observation platforms from which the attack on the fire can be directed by radio.

# Old Reliable Aerial Ladders

One of the greatest inventions in fire fighting was the aerial ladder, which could be carried retracted on a ladder truck and then raised and extended for reaching upper building floors. The modern aerial ladder is raised and then extended hydraulically; but it can also be maneuvered manually in an emergency. It is retracted by controlled gravity. These ladders usually measure about twenty-five feet per section, so that a 100-foot ladder consists of four sections, each one slightly narrower than the one below. New York City's fire department has two 140-foot ladders, but each is only twelve inches wide at the top when extended—a mighty narrow ladder to stand on and work from. The photo above (top) shows an American La France three-section aerial ladder truck of Ladder Company 114, New York Fire Department. This truck is steered from the front, with a tillerman on the rear to help maneuver the rear end around narrow corners. Note the protective roofs over the windshield and the tillerman seat. The photo below shows an aerial ladder in action with the New York Fire Department, with a tower ladder just beyond. An aerial ladder can serve as a water tower, with a ladder pipe nozzle attached to the tip section, or as a stairway for firemen and rescued citizens. On the opposite page two aerials—one used as a watertower and the other as a regular ladder—are at work on a Broadway, New York, fire.

# Fireboats in Action

Many fires around a big waterfront city such as New York are difficult to reach: dock fires, ship fires and waterfront property fires must be battled not only from the land but from the water. Powerful fireboats are the answer, and New York has ten of these floating fire fighters ranging from the Fire Fighter, 134 feet long and weighing 324.17 tons, to a special unit twenty feet long with a single nozzle. The Fire Fighter, and the other units of the Marine Division of the New York Fire Department are stationed at various strategic spots around the waterfront so that they can respond in a matter of minutes. They are equipped with powerful pumps and many monitors or revolving nozzles as shown in these dramatic photos. With the unlimited supply of water below them, they can pour tons of salt water onto a fire for hours at a time. On special occasions their streams are also used as a watery salute to incoming ocean liners or as a welcome to dignitaries visiting the city. These fireboats can not only fight fires with their deck guns, but they can run lines directly to the fire from huge reels on their decks and supply additional water with their pumps.

# NYFD "Scuba Firemen"

When fighting waterfront fires, firemen need to know what is underwater in the area. An obstruction there could prevent a fireboat from approaching the fire closely, an underwater hazard could catch an important hose, or a hard-suction pipe supplying water to a dockside pumper could be obstructed. There could even be a person in the water. For that reason, several waterfront cities have scuba units as part of their fire and/or police departments. Above are some of the underwater firemen of the New York City Fire Department wearing their equipment, with exception of flippers. The two suited firemen in the center are wearing "wet" suits. These suits fit skintight and are made from tough rubber foam with a nylon lining. The thin film of water next to the skin warms up from body heat so the swimmer stays warm even in cold water. The suits on the men with white face masks are called "dry" suits. The suits are entered through openings in the front center. After all air has been pressed out, the suits are sealed shut. Dry suits are best for really icy water, but they can become unmanageable if air gets inside through a puncture or leaky opening. Gloves may or may not be worn with the wet suit. But in the dry suit they are a part of the

whole suit, although in some types they can be removed. The air tanks are carried on the back and can be used for some time underwater before the user has to surface for a refill. The length of time the tank can be used depends on the size of the tank, the activity of the swimmer, and the depth at which he is required to work. In order to stay underwater with the least effort, the men wear weighted belts. The picture above shows the men in the water during a demonstration. They are operating from the *Smoke II*, one of the smaller fireboats. It is fifty-two feet long, 34.5 tonnage, and is powered by two 1,000 horsepower diesel-driven, electric motors. The larger fireboat, the *John H. Glenn*, is seventy feet long, weighs 81.6 tons, and is powered by three 450 horsepower diesels. "Scuba" divers also inspect the hulls and propellers of the fireboats and the intakes for their powerful pumps. The *Smoke II* is what is called a tender boat. Because of its maneuverability and twenty-two miles per hour speed—high for a fireboat—it can carry out many missions that larger and slower fire fighters cannot.

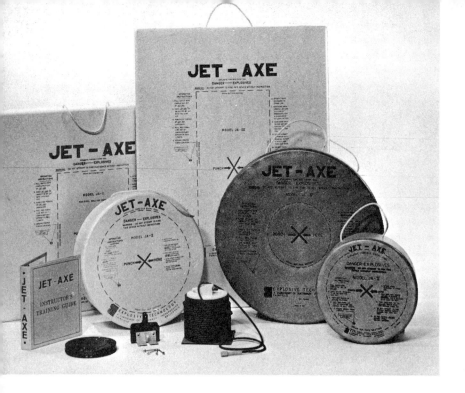

# "Space"-Powered Jet-Axe

One of the most efficient and exciting spin-offs from the space program is a neat little yellow package known as Jet-Axe. In simple language, this is a packaged explosive that directs its full force in one direction only. It can be used to cut neat holes in almost anything in a hurry. It can be detonated in complete safety from a distance and can cut through steel, masonry, ice, reinforced concrete and asphalt roofs, sheet steel, and plastered walls inside a brick wall, with equal ease. Invaluable to fire, police and rescue units, it is a commercial development of the explosive cord used to separate the first and second stages of the Saturn rockets. Developed by Explosive Technology of Fairfield, California, it comes in a variety of sizes, in either a round or rectangular shape. It can either be hung vertically flat against a surface to be opened or placed horizontally flat on a roof or frozen lake. Touched off by an ingenious firing device at the end of a 35-foot cord, the Jet-Axe does in seconds what tools could take a half hour or more to accomplish, particularly if the material to be cut was substantial. On the opposite page, top left to right, we see a fireman hanging a square Jet-Axe against a heavy steel door and then the neat square hole cut a few seconds later. Below this (left) photos show a hole cut in a building roof to ventilate a fire below and a round type Jet-Axe being attached to a heavy reinforced-steel locked door. Bottom photos show (left) the neat hole cut through the steel door, perhaps to rescue fire victims trapped inside, and a similar hole cut through a reinforced concrete wall. The Jet-Axe is a space program spin-off without equal in value when seconds count to save lives or property.

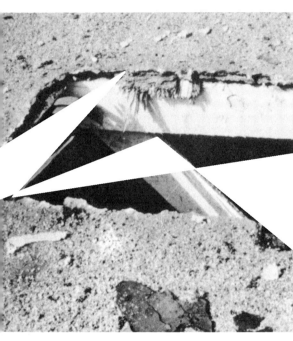

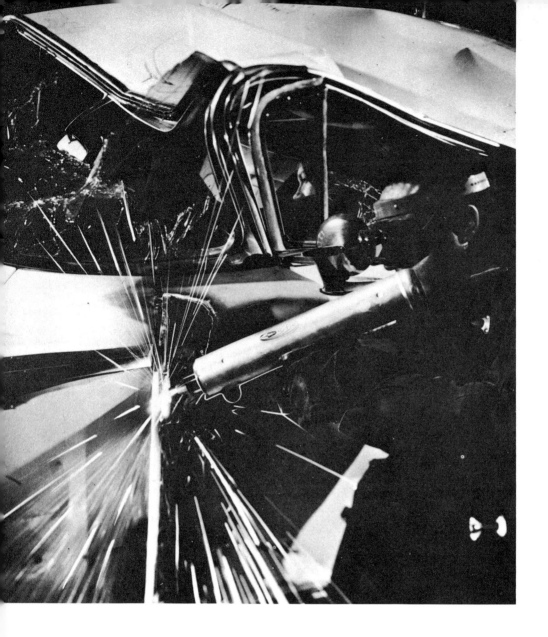

# Rescue Rocket Torch

In many accidents humans are trapped inside crushed vehicles, under girders, or in other metal debris from which they must be cut free quickly. Normally hacksaws, power rescue equipment or cumbersome torches are needed to cut through the metal. Now there is a lightweight, handy, simple-to-use cutting torch that can do the job in minutes. It is a specially designed hybrid rocket motor torch, operated nozzle end forward, which was developed from knowledge gained in work for the space program. The temperature of the rocket flame is more than 5,000 degrees, and it burns for three minutes. This tiny rocket motor torch can cut through ¼-inch steel plate at the rate of a foot a minute and auto steel parts in seconds. The photo shows a fireman wearing safety welding goggles cutting through a car body to free a young woman trapped inside. The lifesaving torch was developed by United Technology Center.

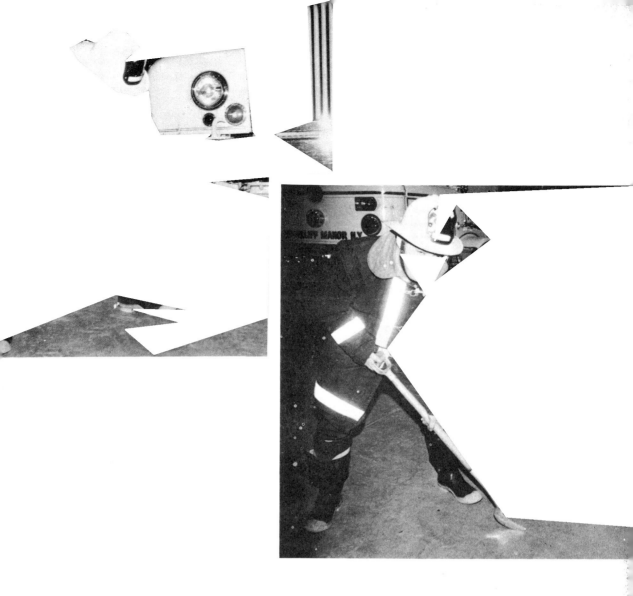

## Multipurpose Saw

It is often important to ventilate or open a roof in order to let out smoke and superheated air from a fire, so that firemen can enter below and get to the flames. Modern roofs are frequently made of a combination of plastic, wood, steel and concrete. This Homelite XL-98 gasoline saw can cut easily through all of these materials. This high-speed saw with special blades can also cut through walls, car bodies, and even concrete floors to reach trapped victims or to introduce water. These photos show the author's son, a ten-year veteran of the Briarcliff Manor Fire Department, demonstrating how to use the saw properly, wearing safety goggles and keeping feet and legs well out of its way. The lower photo shows an older roof "can opener," which is merely a king-sized old-fashioned can opener. These could also be used inside homes for opening up metal ceilings, sheet metal doors, etc. This opener is efficient but slow, and in fighting fires, time is important. Note how the reflecting tape on the fireman's turn-out coat reflects the flash of the camera. These reflective stripes help firemen locate one another at night and in smoke-filled buildings.

# Labor-Saving Water-Vak

Tons of water are usually poured onto a fire, so that even a minor home fire can present a serious cleanup problem. Part of a fireman's job is to help make the dwelling as inhabitable as possible for the victims. In the old days, mopping up flooded floors was a long, backbreaking job. Today, these powerful water vacuum cleaners suck up water quickly and then reverse the flow and eject it from the back tank outdoors, down a drain, or otherwise dispose of it. The author's son (above) demonstrates the operation of a Scott Water-Vak, which can be run from a building's electric current or—if the house current has been destroyed or shut off for safety reasons—from a portable generator carried in one of the trucks. Various types of nozzles can be used in different conditions, and this modern fireman's electric mop can be a real work saver. Directly behind the fireman is Ladder 40 of the Hook & Ladder & Rescue Company, in which I am a life member.

# Lifesaving "Teepee"

This is a most ingenious emergency fire-protection shelter for fighters of brush, grass and forest fires. It could also be used to protect firemen in a sudden wind shift during a city fire. This cone-shaped teepee is made from aluminized and reinforced lightweight material that can be folded to the size of a rolled-up newspaper. In case of danger, the user squats down and slips the unfolded cone over his head; the cone reaches to the ground. The aluminized material reflects heat away from the man inside as the flames sweep past and over him. The teepee is similar in principle to the reflective foil "space-blankets" that reflect body heat back to the user to keep him warm; however, the teepee reflects the heat *away* from his body instead of *toward* it. During a test, the outside temperature was raised to over 600 degrees Fahrenheit, while the inside temperature remained a bearable 130 degrees. This teepee was developed by the U.S. Department of Agriculture's Forest Service with assistance from the Quartermasters Corps.

## Space Age Fire-Safe Fabrics

Developed by NASA, firefighting suits have remarkable qualities that protect the wearer from heat and flames. Here are four men dressed in various types of firefighting equipment. The man on the left is wearing a conventional turnout: coat, helmet and boots. These give him sufficient protection for fighting ordinary fires but not enough for getting close to burning fuels, such as JP-4 jet fuel, or other extremely flammable materials. The other three volunteers from the Houston, Texas, Fire Department are wearing suits designed by NASA for spacecraft crews working in an oxygen-rich atmosphere. The suits are made of special fire- and heat-resistant materials. The helmets worn by the two men on the right can withstand heat well over 1,000 degrees. The visor is made of transparent polysulfone. The photo on the opposite page shows how heat and flame resistant some of these materials really are; they can protect a man's arm from the intense heat of a gas flame held against it. The material being demonstrated is called Beta fabric and is used in insulating fireproof suits. (Cover shows NASA suits being tested.)

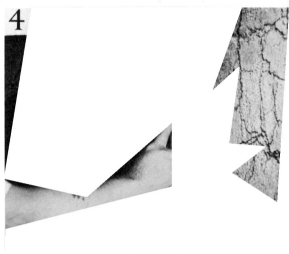

## Space Program Fire Fighters

Fire, one of man's best friends, can also be his worst enemy, especially in the oxygen-rich atmosphere of a spacecraft, where even a spark can be hazardous. Ever conscious of this hazard and of the tragic consequences in the past, NASA's space program sponsors continual research on fire prevention and extinguishing techniques. One important development that has resulted is a fire-retarding paint. Above, a test shows how amazingly effective this paint is. The test structure on the left was unpainted, while the one on the right was painted with the new paint developed at NASA's Ames Research Center in California. This paint enlarges with the heat and chars. The heavy charring serves as an insulation which protects the inner layers of the paint and the wood beneath it. Three coats of paint were applied to the structure on the right. Note that in photo 4 some of the original paint layers on the structure are still intact and so is the wood. In photo 2 the fire has gone out in the right-hand structure; in photo 3 you can see the effect of the raging fire on the untreated and treated structures. This type of paint will save many lives when it is put into general use in private dwellings, motels and hotels.

## Fireproof and Germproof

Fireproofing can also be applied to vehicles, aircraft and other structures. The photo above shows the interior of the Mobile Quarantine Facility used during the first of the Apollo moon missions. The floor, panels and drapes were sprayed with nonflammable fluorel, the seats covered with fire-resistant synthetic modacrylic material, and even the seat belts were made of a nonflammable webbing. Eventually all vehicles may be as fireproof as this special quarantine van designed to safeguard our returning moon men.

# Machines Against Forest Fires

Fighting forest and brush fires has always been an important part of fire fighting. Today, especially with our new concern for the environment, it is even more important. Every year fires destroy not only millions of acres of forest timber, but the habitat of wildlife as well. The burned-over lands are exposed to erosion, the streams in the area are polluted, and so is the atmosphere. Here are some machines used by the United States Forest Service to fight the battle. The photo above shows a sturdy four-wheel-drive fire truck by International, designed for fighting forest fires. It is a pumper that carries 300 gallons of water and can carry a fire crew both inside the cab and behind it. The compartments can carry equipment such as grass-fire brooms, Indian Tanks (hand-operated backpacks filled with water), chain saws, axes, brush hooks, shovels and other fire-fighting gear. Note the preconnected hand line on a reel at the back of the truck, and the radio antenna. Instant communication is all important in fire fighting. On the opposite page (top) a couple of one-man fire-line-digging tractors are being demonstrated before a group of fire school students. These tractors can do the work of several men in clearing a fire break ahead of a fire, which stops its progress along the ground. The lower photo shows a powerful Cat, short for caterpillar, bulldozing a fire break ahead of burning ground debris and brush. Often as many as thirty of these caterpillar tractor bulldozers are used to contain a bad fire.

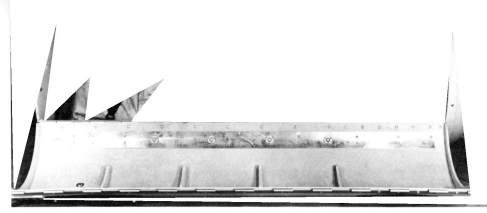

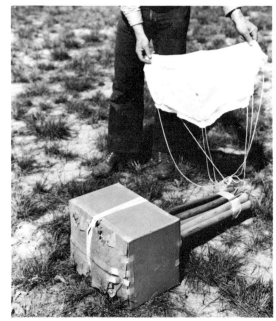

# Crash Pads for Firemen's Tools

Delivering tools to forest-fire fighters deep in the wilderness is quite a problem, but the U.S. Forest Service neatly solves it with an ingenious use of honeycomb paper. The honeycomb paper is packed in the bottom of tool cartons and drop capsules so that when the landing occurs the paper honeycomb crushes, absorbing most of the shock of the landing. The tools or equipment are left undamaged. The top photo shows a fiberglass wing cargo capsule containing a chain saw with accessories. Note the crushable honeycomb paper at right end; this becomes the bottom when the capsule is dropped. Below are two photos of a package of Pulaski tools ready to be dropped (left), and after being dropped (right) with no damage to the tools. The small parachute attached to the tool package is there to stabilize the drop rather than retard the fall to any extent. A similar chute can be attached to the chain saw canister. In *Chute I* describe the use of these honeycomb shock absorbers for dropping military vehicles and weapons via parachute. Thanks to space-age developments, all sorts of supplies can be air-dropped to fire fighters.

# Science vs. Flames

One of the frustrations of planning strategy against a forest fire is the dense cover of smoke that makes aerial observation nearly or completely impossible in many cases. Space-age science has now made penetration of this smoke cover possible. The photo above (top) shows the console of a T-29B aircraft flying over a smoke-concealed fire area. Although the fire itself can not be seen, its presence is instantly revealed by red lights flashing before the observer. In the lower photo is a piece of a film taken by a new system called Fire Scan. This photo was taken from an infrared screen and shows a whole forest fire area. The white areas show where the fire is hottest, and the darker areas where the fire has either burned over or has not yet touched. Although the flames are completely concealed by smoke, thanks to this revealing photo, flame-extinguishing chemicals can now be accurately dropped on the fire itself instead of being wasted on burned-over land or untouched forest. Note also the details of the terrain, streams, slopes and dammed reservoir, that are revealed by the Fire-Scan system.

# Flying "Fire Trucks"

Aircraft of many types ranging from small observation planes to helicopters and special-purpose aircraft, fill important roles in forest-fire fighting. Here are some of the ingenious ways they are used to fight fire. In the photo above, a Bell helicopter lays fire hose from the air. Hose streams out behind the chopper far faster than it could be dragged by men on the ground. One end of the hose is already attached to a pumper and the nozzle end will be dropped near the men on the fireline in a matter of minutes. In the top photo on the opposite page, a 205 chopper lifts a filled water bucket from a stream. It will carry the water to the fire scene and dump it on the flames, or else soak vegetation ahead of the fire. The center photo shows a larger S-61 helicopter carrying an ever larger "fire bucket" to a blaze. And the lower photo shows an air tanker aircraft hurling clouds of fire-smothering borate on a fire below. This kind of aerial fire fighting takes great skill and courage, but it is very effective.

## Firemen in Jumpsuits

In all fire fighting, the men are the final all-important weapon against the flames. They wield the tools, drive the machines, fly the aircraft, and plan the strategy. "Smokejumpers" of the U.S. Forest Service are highly skilled and courageous fire fighters who parachute into the fire zone to stop the fire while it is still small and possible to combat with minimum equipment. Here are three "smokejumpers" dressed for a drop. Note the steel "cages" that will be lowered over their faces for protection from branches, the main chutes on their backs and the reserve chutes worn in front. High collars and strong gloves give added protection. The bulging lower leg pockets contain safety line to help them get down from any trees they may land in. On the opposite page, two jumpsuited firemen start down into the forest. Note the white parachutes caught in trees from earlier jumpers who have made treetop landings.

# Down, and Up a Tree!

In these two dramatic photos from the U.S. Forest Service you can see the hazards forest-fire fighters face when they bail out into a fire area. In the photo above you can count many parachute canopies hanging in tall trees. On the opposite page you can see how a jumper gets down. He attaches one end of his descending line to his parachute harness or to another firm anchor and then uses it to work his way down through the branches to the ground. This technique enables the smokejumper to get into action on the fireline, and if the wind shifts and the fire roars in his direction, he is not left dangling helplessly in its path, unable to descend and out of reach of rescue. Courage, strength and training all contribute to successful use of the smokejumper's equipment. Like all fire fighters, smokejumpers must be willing to take risks.

## What It's All About?

In the end the purpose of all fire fighting is to protect property and save lives. This fine photo by Fireman H. Casazza, of the New York City Fire Department, shows fellow members of the largest fire department in the world as they help two rescued fire victims down a 100-foot aerial ladder to safety. No matter where the fire occurs or what type it may be, volunteer or professional fire fighters are ready and trained to go into action, to use the latest in fire-fighting weapons with skill and courage. If there is a firehouse near you, visit it and become acquainted with the men and their equipment. It could be the most interesting and important visit you ever made. Thanks to space age materials, new training techniques, and ingenious new weapons, our fire fighters can keep pace with today's new fire hazards.